WEIRD SEA CREATURES™

THE SEA SLUG

Nudibranchs

Miriam J. Gross

The Rosen Publishing Group's
PowerKids Press™
New York

For Martin, who is always helpful

Published in 2006 by The Rosen Publishing Group, Inc.
29 East 21st Street, New York, NY 10010

First Edition

Editor: Daryl Heller
Book Design: Albert B. Hanner

Photo Credits: Cover © Yoshi Hirata/SeaPics.com; pp. 5, 6, 9, 21 © Doug Perrine/SeaPics.com; p. 10 © Robert Yin/Corbis; p. 13 © Jeff Jaskolski/SeaPics.com; p. 14 © Marc Chamberlain/SeaPics.com; p. 17 © James D. Watt/SeaPics.com; p. 18 © Carlos Villoch/V&W/SeaPics.com.

Library of Congress Cataloging-in-Publication Data

Gross, Miriam J.
  The sea slug : Nudibranchs / Miriam J. Gross.— 1st ed.
     p. cm. — (Weird sea creatures)
  Includes index.
  ISBN 1-4042-3191-9 (lib. bdg.)
  1. Nudibranchia—Juvenile literature. I. Title.
  QL430.4.G76 2006
  594'.34—dc22
                              2005000728

Manufactured in the United States of America

Contents

GEMS OF THE SEA

Sea slugs have poor eyesight. They depend on rhinophores, which grow from their heads. Sea slugs use rhinophores to find prey and sense other objects in the water. The shape, size, and color of the rhinophores are not the same on different species of sea slugs.

Sea slugs are among the most beautiful animals in the ocean. There are about 5,000 **species** of sea slugs. They come in a wide range of colors and shapes. Their colors include blue, yellow, pink, and green. Sea Slugs can be spotted or striped. Some have feathery branches growing from their backs. Others are covered with rows of tiny tubes. Many are less than 1 inch (2.5 cm), but some can grow to be more than 1 foot (.3 m) long. Sea slugs live on the ocean floor throughout the world.

Sea slugs have no shell, so these soft, wormlike creatures could be an easy catch for hungry sea creatures. However, most other animals avoid sea slugs. Their bright colors and patterns warn other animals that sea slugs are **toxic**. Many poisonous animals are brightly colored. Not all sea slugs are toxic, but their bright colors let other animals know that they may be a danger.

This sea slug is a Nembrotha kubaryana. The orange hornlike growths on the animal's head are rhinophores. The rhinophores are sensors, which help the sea slug smell and find food in the water. A Nembrotha kubaryana is a large sea slug that can grow to about 4 inches (10   in length.

This sea slug is a *Reticulidia halgerda. Sea slugs use their foot, which is usually on the underside of their body, to move. The movement occurs when the foot contracts and then relaxes, or pulls together and then lets go and spreads wide. The creature pulls its body forward a little with each contraction. A sea slug's body also produces mucus, which can be sticky or slippery. This helps the creature stay put or slide.*

SEA SLUGS AND THEIR RELATIVES

Scientists separate the animal kingdom into many groups. This helps them study how different creatures are related to each other and how they have **evolved**. Sea slugs belong to a large group of creatures known as mollusks. Mollusks are soft-bodied invertebrates, or animals without backbones. This group includes slugs, snails, octopuses, squid, clams, and mussels.

Sea slugs are part of a smaller group within mollusks called gastropods. The name gastropod means "stomach-foot," because of the shape of the animal's body. The body of most gastropods is made up of a small head and a mantle. The mantle contains the stomach, **digestive glands**, **intestine**, kidney, heart, and **reproductive organs**. The foot is found beneath the body. Sometimes the mantle makes it hard to see the foot. The common slugs and snails you might see in your garden belong to this group, as do the many other kinds of slugs and snails that live in the ocean.

NAKED GILLS

The scientific name for the sea slugs in this book is nudibranch (pronounced "NOO-dih-brahnk"). Nudibranch means "naked gills." Gills are **membranes** that take **oxygen** from the water. They allow fish and most other sea creatures to breathe underwater. Most other sea creatures' gills are inside their bodies. The gills of sea slugs, however, stick out from their bodies into the water. You can recognize the different groups of sea slugs by the types of gills they have.

In the group of sea slugs known as Dorids, the gills take the form of bushy, featherlike bunches at the sea slug's rear end. In the Aeolid group of sea slugs, the gills are part of the thin tubes that cover the sea slug's back. These tubes are called cerata (pronounced "ser-AH-tuh").

A sea slug's gills are often brightly colored. They add to the creature's lovely and strange appearance.

This is a nudibranch in the Aeolid group. The white curving growths that cover this creature are cerata. These cerata help this creature breathe. The purple coral this sea slug is eating is called a sea fan, or a Gorgonian.

9

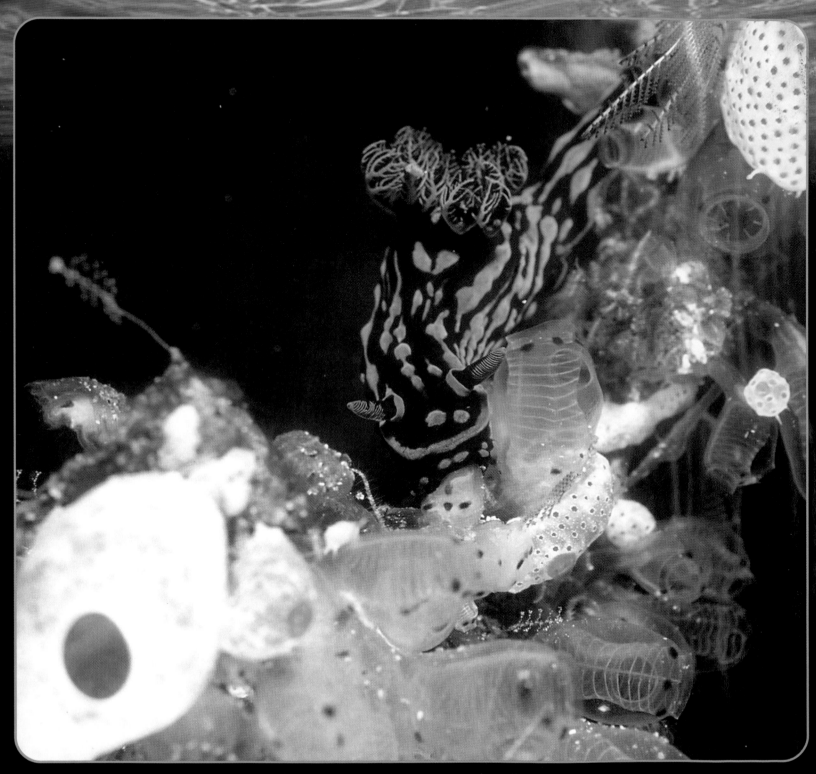

This sea slug, which is a Dorid, is eating small clear animals called tunicates. Tunicates are sometimes called sea squirts because they squirt, or spray, out a stream of water to keep themselves safe when they are in danger.

YOU ARE WHAT YOU EA

All sea slugs are carnivores, which means they eat animals. They eat **sponges**, jellyfish, **sea anemones**, corals, or other sea slugs. E species usually eats only one kind of anima. Their lifestyle, their body, and even the shape of their radula, which is like a tongue covered with tiny teeth, have evolved to help them eat that animal. The radula comes out of the sea slug's mouth and **scrapes** or bites food.

Some Dorid sea slugs, such as the small, red *Rostanga pulchra*, eat sponges and will spend their life on top of a sponge. The *Rostanga pulchra*'s body is red to match the color of the sponge it eats. Other animals will not eat sponges because of bad-tasting or poisonous **chemicals** in the sponge's body. These chemicals are not harmful to Dorids, however. A Dorid can even store these chemicals in its own body to make itself poisonous to **predators**.

however. When an Aeolid, such as the flame nudibranch, eats a sea anemone, it passes the stinging cells to the ends of the cerata. If attacked the Aeolid can throw off a few cerata and sting, or stab, the attacker. New cerata grow back in a few weeks.

...NG FOOD FROM THE SUN

...e species of Aeolid sea slugs have a **symbiotic** ...tionship with the algae that live inside their body. ...lgae are plantlike living things without roots or stems, which live in water. The blue dragon, a sea slug found near southeastern Australia, takes in algae when it eats other animals in which the algae are already living. The blue dragon passes the algae into its skin, which is so thin that sunlight can pass through it.

The algae living in the blue dragon's skin produce food through photosynthesis, the method that plants use to get power from the Sun. The algae take in the Sun's light and turn it into sugars. The blue dragon lives off the sugars being produced within its own body.

Phyllodesmium longicirrum is another sea slug that uses algae to help produce food for itself. It is found in Indonesia. This creature's body is covered in brown rings, which are actually gardens of algae living under its skin.

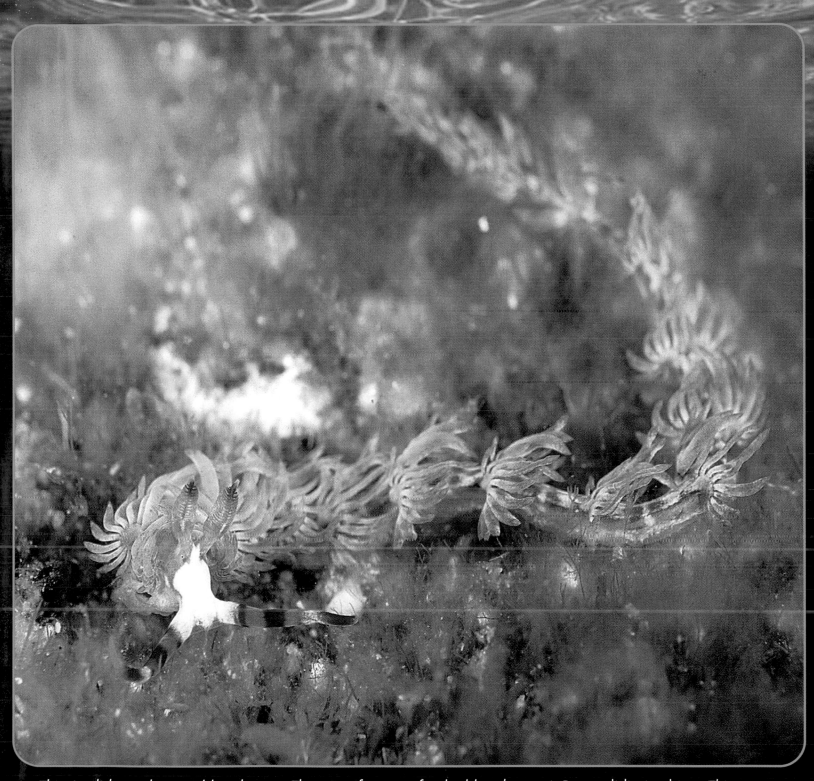

This Aeolid sea slug is a blue dragon. The scientific name for the blue dragon is Pteraeolidia ianthina. This blue dragon gets much of its food and color from the algae that live inside its body. The algae, in turn, get their food by turning sunlight into sugar.

This Aeolid sea slug was found in the Pacific waters off San Miguel Island. San Miguel is about 55 miles (88.5 km) west of the Southern California coast. The island is home to many other sea creatures, including sea lions and seals.

THE WORLD OF A SEA SLUG

Sea slugs live all over the world, but the greatest number of them live in the warm **tropical** waters around the Caribbean Islands, Indonesia, and Australia. Most sea slugs live in water that is not very deep. They crawl along the ocean floor or on algae with their single foot. A few species live deeper in the ocean, however. The sea slugs of the deep sea swim by slowly waving their soft bodies through the water.

The Spanish dancer is one of the largest sea slugs and can grow up to 1 foot (.3 m) long. Instead of crawling on the ground as most sea slugs do, the Spanish dancer swims by slowly waving its red, orange, and pink body through the water.

Sea slugs are often found in coral reefs, which are colorful areas near the shore that are home to many different animals. These animals include brightly colored fish, giant clams, sea stars, giant sea turtles, sea urchins, and moray eels. The reefs are underwater hills made from the external, or outer, **skeletons** of coral polyps. A coral polyp is a soft, tubular animal with a central mouth that is surrounded by tentacles, or long thin growths.

BRIGHT COLORS FOR SAFETY

The toxic chemicals in some sea slugs' skin do not have an effect on the sea spider. The sea spider hunts sea slugs while they crawl on algae. The sea spider will hold the sea slug still with its front claws and then eat it whole. Some species of sea slugs eat other sea slugs.

Many mollusks, such as clams and snails, have a shell. Sea slugs, however, have no shell. This allows them to move more freely, since they do not have to carry a heavy shell. Having no shell, however, leaves the sea slug's soft body without the **protective** cover that keeps other mollusks safe. Therefore, sea slugs use their coloring to avoid predators. Some sea slugs are the same color as their surroundings and cannot be easily seen by predators. For example, a purple sea slug in Australia called the Jorunna lives on a purple sponge.

In nature bright colors may sometimes be a sign that an animal is poisonous or bad tasting. Some sea slugs, such as the Chromodorids, have bright colors and patterns that make them stand out in their surroundings. These colors make animals think that the Chromodorids are not good to eat. This is called defensive coloration.

cerata

rhinophore

rhinophore

mantle

This Spanish shawl nudibranch has the scientific name Flabellinopsis iodinea. The purple, orange, and red colors on its body are caused by the orange animals this creature eats. A Flabellinopsis iodinea can escape a predator by swimming away. This sea slug swims by bending its body from side to side.

Because sea slugs are both male and female at the same time, it is easier for them to find another sea slug to mate with. Plants and animals that do not have a backbone often have these mating habits.

How Sea Slugs Mate

Sea slugs are simultaneous hermaphrodites, which means they are both male and female at the same time. This makes it easier for sea slugs to **reproduce**, since any other member of the species can be a **mate**. Sea slugs mate in pairs. Both animals **fertilize** one another's eggs at the same time. Therefore, both creatures will become mothers. They cannot fertilize their own eggs, however.

Sea slugs lay their eggs in masses that number from hundreds to millions. They wrap the eggs in mucus, which is a thick, sticky liquid, to keep them together and attach them to the ocean floor. The egg masses are usually white. They can also be brightly colored to match the animal on which the sea slug lives and feeds.

Many of the sea slugs that hatch, or break out of, these eggs will not live to adulthood. Predators will eat them. Having so many babies at the same time makes it more likely that at least some of them will live to reproduce.

LIFE CYCLE

Most sea slugs hatch from their eggs as veliger larvae, which are free-swimming creatures that have a clear, thin shell. Predators will eat many of these tiny larvae.

The larva will go through a change once it finds the food that its species eats. The creature then settles onto the ground and leaves its thin shell behind. At this stage it is a smaller **version** of the crawling slug it will become as an adult. These tiny sea slugs will continue to grow until they reach maturity, or the age at which they can reproduce. The time this takes depends on the species, as well as how hot or cold the water is, and the supply of food.

Most species of sea slugs live for about one year, but some species have an even shorter life. A tropical Aeolid called the *Tenellia pallida* reaches maturity in around three weeks and usually dies within two or three months.

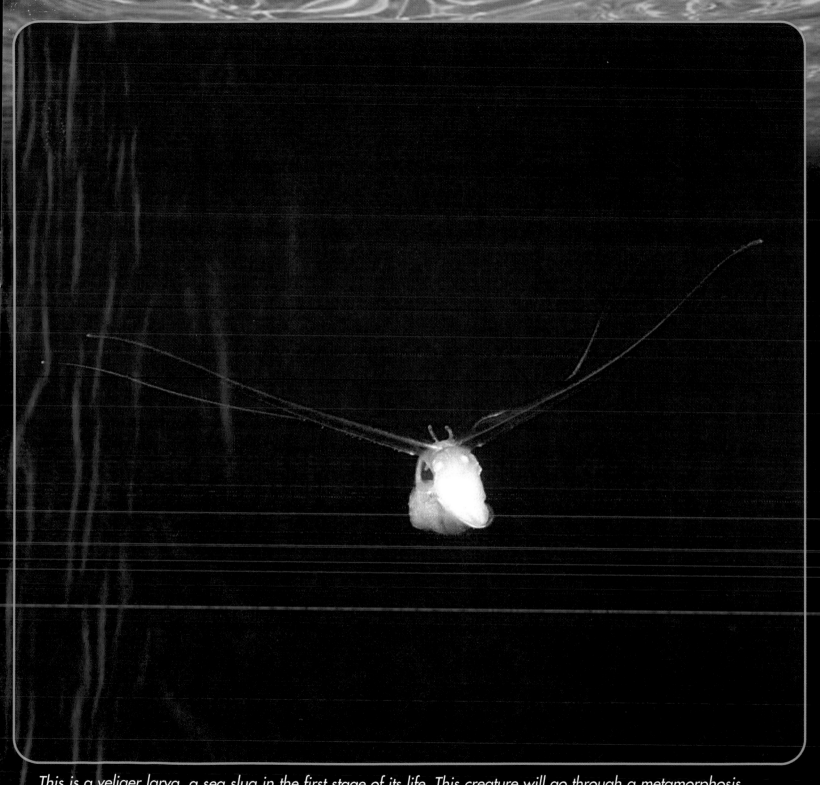

This is a veliger larva, a sea slug in the first stage of its life. This creature will go through a metamorphosis, or completely change how it looks, and become a sea slug that crawls on the ocean floor.

SEA SLUGS AND SCIENCE

Sea slugs are helpful to people, as well as being beautiful. Scientists are always looking for new ways to treat illnesses. They have recently discovered that some chemicals found in sea slugs can be used as **medicines**.

Sea slugs such as the *Jorunna funebris* have a chemical in them called mimosamycin that can be used as an antibiotic. An antibiotic is a medicine that kills microorganisms, or tiny creatures that can make you sick. Mimosamycin is especially useful against mycobacteria, the kind of **germs** that cause the lung **disease** tuberculosis.

From their special chemicals to their bright colors, sea slugs are a good example of the weird and wonderful ways in which animals evolve. This creature's unusual shape and bright colors, and its ability to take on the defenses of its **prey**, have allowed the sea slug to stay alive in its ocean home.

GLOSSARY

chemicals (KEH-mih-kulz) Matter that can be mixed with other matter to cause changes.

digestive glands (dy-JES-tiv GLANDZ) Body parts that help break down food into energy.

disease (duh-ZEEZ) An illness or sickness.

evolved (ih-VOLVD) Changed over many years.

fertilize (FUR-tuh-lyz) To put male cells inside an egg to make babies.

germs (JERMZ) Tiny living things that can cause sickness.

intestine (in-TES-tin) The part of the digestive system that is below the stomach.

mate (MAYT) A partner for making babies.

medicines (MEH-duh-sinz) Drugs that a doctor gives you to help fight illness.

membranes (MEM-braynz) Soft, thin layers of living matter that come from a plant or an animal.

oxygen (OK-sih-jen) A gas that has no color, taste, or odor, and is necessary for people and animals to breathe.

predators (PREH-duh-terz) Animal that kill other animals for food.

prey (PRAY) An animal that is hunted by another animal for food.

protective (pruh-TEKT-iv) Keeping from harm.

reproduce (ree-pruh-DOOS) To have babies.

reproductive organs (ree-pruh-DUK-tiv OR-genz) The body parts inside an animal that allow it to make babies.

scrape (SKRAYP) To rub something off by using force.

sea anemones (SEE uh-NEH-muh-neez) Soft, brightly colored sea animals that look like flowers.

skeletons (SKEH-lih-tunz) The bones in the bodies of animals or people.

species (SPEE-sheez) A single kind of living thing. All people are one species.

sponges (SPUNJ-ez) Sea animals made of hard or soft matter that allows water to pass through it.

symbiotic relationship (sim-bee-AH-tik rih-LAY-shun-ship) When two living things use each other to live and are helped by this tie. However, one living thing may be helped more by this tie.

toxic (TOK-sik) Poisonous.

tropical (TRAH-puh-kul) Having to do with the warm parts of Earth that are near the equator.

version (VER-zhun) Something different from something else, or having a different form.

INDEX

A
Aeolid, 8, 12, 20

B
blue dragon, 12

C
cerata, 8
chromodorids, 16

D
digestive glands, 7
Dorid(s), 8, 11

G
gastropod(s), 7

I
intestine, 7

M
mollusks, 7, 16

N
nudibranch, 8

P
Phyllodesmium longicirrum,
     12
predators, 11, 16, 19–20

R
reproductive organs, 7
Rostanga pulchra, 11

S
sea anemones, 11
sponges, 11
symbiotic relationship, 12

T
Tenellia pallida, 20

V
veliger larvae, 20

WEB SITES

Due to the changing nature of Internet links, PowerKids Press has developed an online list of Web sites related to the subject of this book. This site is updated regularly. Please use this link to access the list:

www.powerkidslinks.com/wsc/seaslug/